MY FAVORITE COLORS

by Pearl Markovics

Consultant:
Beth Gambro
Reading Specialist
Yorkville, Illinois

Contents

My Favorite Colors 2

Key Words 16

Index 16

About the Author 16

New York, New York

I love colors!

I love blue.

I love red.

I love green.

Grass is green.

Pigs are pink.

Pick your favorite!

What color do you love?

Key Words

blue

green

pink

red

yellow

Index

blue 4–5   pink 12–13   yellow 10–11
green 8–9   red 6–7

About the Author

Pearl Markovics has many favorite colors. She especially likes the colors midnight blue and bright orange.